SCIENCE FAIR
HOW TO DO A SUCCESSFUL PROJECT

CHEMISTRY

BY
SHIRLEY COX

SERIES CONSULTANT
DR. JOHN M. LAMMERT
Associate Professor of Biology
Gustavus Adolphus College
St. Peter, Minnesota

ROURKE
PUBLICATIONS
INC.
Vero Beach, FL 32964
U.S.A.

LIBRARY OF CONGRESS CATALOGUING-IN-PUBLICATION DATA

Cox, Shirley 1953–
Chemistry / by Shirley Cox
p. cm. — (Science fair)
Summary: Explains the scientific method and suggests a variety of chemistry projects and experiments suitable for a science fair.
ISBN 0-86625-427-7
1. Science—Experiments—Juvenile literature. 2. Science projects—Juvenile literature. 3. Science—Exhibitions—Juvenile literature. 4. Science—Experiments—Methodology—Juvenile literature. 5. Chemistry—Experiments—Juvenile literature
[1. Chemistry—Experiments. 2. Science projects. 3. Experiments 4. Science—Experiments—Methodology]
I. Title II. Series. 92-6974
Q164.C73 1993 CIP
540' .78—dc20 AC

DESIGNED & PRODUCED BY:
MARK E. AHLSTROM
(The Bookworks)

PHOTOGRAPHY:
Cover–THE IMAGE BANK/Filsner
Text–MARK E. AHLSTROM

TABLE OF CONTENTS

CHAPTER 1

What Makes A Science Project "Scientific?"

How does the idea of doing a science fair project make you feel? If you decided to do one yourself, you are probably very excited. If your teacher has assigned a project as a class requirement, you may not feel quite so happy about it. You may even be feeling a little afraid.

Once you understand how to use the **scientific method**, your fear should disappear. It will be replaced with the excitement of doing something challenging and fun. Why should this miracle occur? Because there are no wrong ideas, wrong experiments, or wrong **observations**! In fact, you can't be wrong. You can start with any idea. The scientific method will help you choose a reasonable idea. If you make errors, you don't even have to know it! The scientific method will help you pinpoint mistakes and help you decide what changes to make.

When you're using the scientific method, even the "wrong" results can be "good" results. Take the example of Dr. Spencer Silver.

Dr. Silver worked in product development at the 3M company. Many years ago he had the problem of developing a strong, permanent glue. Using the scientific method, he planned and conducted an experiment. When Dr. Silver tested the product that was formed, he found that it wasn't a strong glue at all. It was barely sticky and not at all permanent. The experiment was a failure and Dr. Silver had to try again.

Several years later, another scientist at the 3M company, Art Fry, was upset. He was always losing his notes and page markers in a book he was using. Then he thought of Dr. Silver's experiment. Mr. Fry used the barely-sticky glue to create the product sold today as self-stick notes!

Self-stick notes were discovered by accident.

So what is this scientific method?

The scientific method is a way to answer questions and solve problems. It involves a series of steps that must be followed in a logical order. Scientists use it to find explanations for things they have observed. When scientists notice something new or interesting, they wonder "What is that? Why did that happen?" A problem has been identified. Now the scientist will form a **hypothesis** to explain the observation. A hypothesis is a possible answer that is based on some knowledge of the problem. This knowledge comes from finding out what other scientists have already learned about similar problems.

Not all scientists agree on the order of the steps when using the scientific method. Different problems often require a different order. However, scientists do agree on what the necessary steps are. A typical sequence is shown in the chart below. As long as all the steps are included and followed in a logical order, the scientific method will work for you.

Steps Followed in the Scientific Method

- ✔ Make an observation.
- ✔ State the problem: What do you want to find out?
- ✔ What is already known about the observation?
- ✔ Develop a hypothesis: What do you think is a reasonable explanation for the observation?
- ✔ Design an experiment that will provide answers: What materials will be needed and how will they be used.
- ✔ Record data or observations: What happened during the experiment?
- ✔ Analyze the results.
- ✔ Draw a conclusion: Did your data support your hypothesis? What did you learn? What do your results mean?

Now let's look at an example of an observation and a hypothesis. Have you ever used hydrogen peroxide as an antiseptic on a cut? Did you see bubbles form? The liquid fizzes as the result of a **chemical reaction** which breaks down the hydrogen peroxide into new substances. One **product** of the reaction is oxygen, a gas that makes the bubbles. This breakdown happens in a stored bottle of hydrogen peroxide, too, but it is so slow that human senses can't detect the release of oxygen. Sooner or later you have to throw away the bottle because it doesn't "work" anymore. You might begin to think that blood, or something in blood, speeds up the chemical reaction enough that you can see bubbles form. Perhaps you wonder if other materials could also speed up the reaction. Good for you! You are thinking like a scientist. **Organic matter** comes from living things, while **inorganic matter** does not. Blood is organic matter. Your hypothesis might be "Organic materials can speed up the reaction of hydrogen peroxide, but inorganic ones do not."

Next, a scientist makes a plan to test the hypothesis. This plan is called the **experimental design**. It involves choosing materials to test the hypothesis and very specific steps for their use in the experiment. *Each step must be carefully looked at to make sure it is safe.*

To test our hypothesis, items to test must be chosen. Organic materials could include potato and liver. Inorganic items could include salt and aluminum foil. Plastic medicine cups can be used to hold the items and the hydrogen peroxide. A fifth cup of hydrogen peroxide, to which no item is added, is called the **control**. The control makes sure that nothing besides the test items causes bubbles to form.

When all the materials are ready and the cups are labelled, the actual experiment can begin. Each test item is added to a cup one at a time.

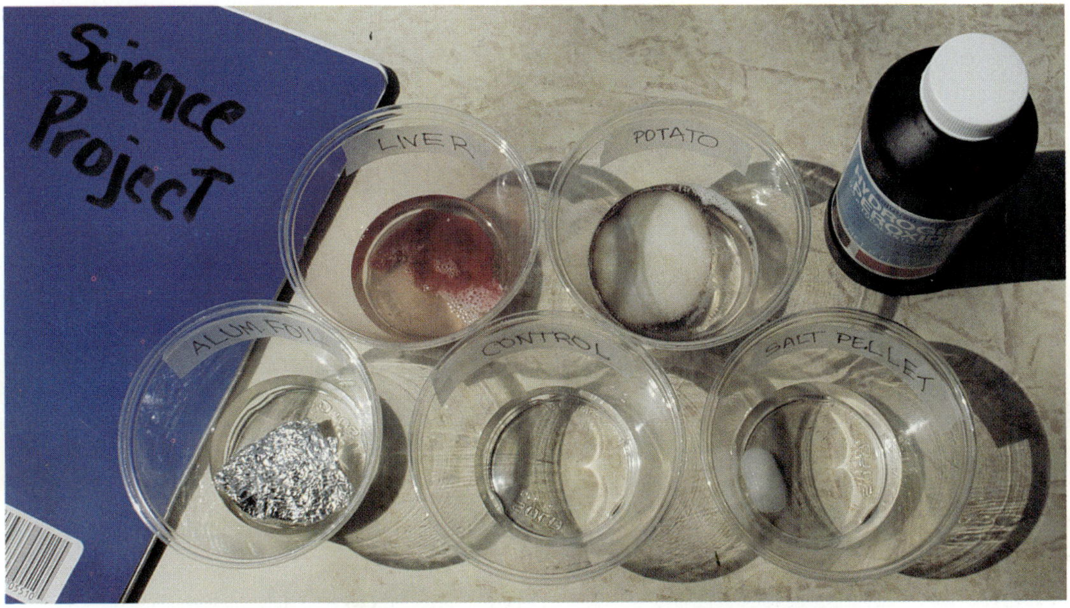

It's important to have a control in your experiment.

During the experiment **data** are collected. This means that what is seen or measured is recorded. In our example, the formation of oxygen bubbles in each cup is observed. The appearance of the hydrogen peroxide and each test item is noted before and after the reaction. Our hypothesis predicts that oxygen will be produced in the cups to which potato and liver are added, but not in the other three cups.

After this first trial of the experiment, the results are studied. In which cups did the reaction occur? Were there any differences in the speed of the reactions? Were there any changes in the test items?

Finally, from these results, **conclusions** are made. Was the hypothesis correct? If not, what would be another hypothesis? Should the experiment be redesigned or repeated before the hypothesis is accepted or rejected?

The data produced in an experiment often lead to new questions and more experiments. Many times observations made during an experiment suggest new problems that require more work. Even the work of a Nobel Prize-winning scientist produces more questions than answers.

Scientists like to tackle new problems.

CHAPTER 2

Choosing Chemistry As A Topic For Your Project

Many chemical processes occur in the world around you, and even inside your body. Studying chemistry can be very exciting. Chemical reactions can form the basis of a good science fair project.

What makes a successful chemistry project? The hardest part is picking a topic. Spending some time to think about a number of ideas will help insure the success of your final choice. The best topics will be those that interest you. You will be excited to design and try out these experiments. Your science project will no longer be a task assigned by your teacher, but your own creative idea. Your idea doesn't have to be complex or hard to do, especially if this is your first project. Great projects often focus on something simple.

When doing a chemistry project, safety must always be a big concern. Chemical reactions require the mixing of **reactants** to produce new substances. Such changes can release **toxic** fumes or large amounts of heat. The new substances might damage or destroy the container. Before any work is actually begun, you need to find an adult who understands chemistry to be your supervisor. Your supervisor will check your experimental design to make sure that it is safe. Then you must follow the agreed-upon **procedure** exactly. If you want to try anything different, your supervisor must approve the changes, too.

This book will guide you through the basic steps you need to follow in order to do a good science fair project about chemistry. You will not find step-by-step instructions. Instead, you will discover ways to use your own imagination and creative powers. Projects that you create will be fun to do!

Types of Projects

If you could visit a science fair to look at the projects, you would probably see three different types:

- ★ *Exhibits* display collections or explain a topic.
- ★ *Demonstrations* repeat an idea or experiment found in a book.
- ★ *Investigations* use the scientific method to solve a problem.

Because of the way the projects are judged, the first two types often receive lower scores and fewer awards. They are less scientific than an investigation. The rules of some fairs may even prohibit any project that is not an investigation. Most science fairs will not allow projects that are done by groups of students, either. You should find out how your project will be judged. You need to know if there are any restrictions on the type of projects before you decide what you want to do. Perhaps your teacher can review your plan and help you find out if it meets the fair's rules.

If all your ideas seem to be exhibits or demonstrations, don't panic. The next section of this book shows how these projects can be expanded and changed into an investigation. In fact, because chemistry can be dangerous, even chemists usually start with an experiment that has already been done.

☞ *Example 1*

Perhaps you have seen models of atoms or compounds that are made of colored balls and sticks. Maybe you are thinking about building a display of such models. This plan can result in a nice exhibit, but with a little more work it can be made into a demonstration. During a chemical reaction, some **chemical bonds** may be broken and new ones formed. You might build models of these **reactant molecules**, too. This would change your exhibit into a demonstration by showing how bonds break and reform. You might also show the reaction with the actual compounds. Your models demonstrate what actually occurs during a reaction, even if it cannot be seen.

Models show the make-up of a compound.

Models can also be used in an investigation. Scientists often use computers to build and then change models of molecules in order to predict how they might behave in a chemical reaction. A very famous use of models helped two scientists win a Nobel Prize. James Watson and Francis Crick fit together model pieces to determine the structure of **DNA**. DNA is a very large molecule found in living things that contains the genetic information that parents pass on to their offspring. However, using models in an investigation requires a good knowledge of chemistry. This kind of investigation may be too difficult for most students.

☞ *Example 2*

Have you seen ads on television about pH-balanced shampoo, antacid tablets that get rid of stomach acid faster, or buffered aspirin that won't upset your stomach? All of these advertisements refer to acid-base chemistry. You could collect examples of **acids** and **bases** and put them into an exhibit. Your exhibit might explain how acids and bases differ.

With some more study, you could find out how chemists use the **pH scale** to measure the strengths of acids and bases. Then you could measure the pH of each sample in your collection. Now your project is a demonstration. Why not use your creativity, work a little harder, and design your own investigation? For example, find out what a **buffer** is. Then use your new knowledge to design an experiment to find out which aspirin buffers best. Find out how acids can be neutralized, then test antacids to find out which one really absorbs the most stomach acid. Study about the structure of hair, then do an experiment to see if a pH-balanced shampoo really can repair damaged hair. All three ideas would be good investigations.

☞ *Example 3*

An exciting area of new research is polymer chemistry. **Polymers** are giant molecules made from smaller, simpler units that repeat themselves in a pattern. Natural polymers include paper, cotton, silk, as well as molecules like protein and carbohydrates in the food you eat. Nylon, polyester, and plastics are man-made polymers. You could collect samples of polymers and build an exhibit showing their many uses.

Most fabrics are made from polymers.

As you read about polymers, you may find some recipes for making some yourself. Silly putty is a polymer you can make by mixing a **saturated solution** of borax with Elmer's glue. Now your polymer project can become a demonstration of how polymers are made.

However, this demonstration can form the basis for many interesting investigations. How will the polymers be changed if you use more or less borax solution? What will happen if you dilute the glue with water before you add borax? Can you color the polymer by adding food coloring? What is the best mixture of glue and borax to produce a polymer with the highest bounce, or the greatest stretch, or the fastest flow? Does the temperature change during the formation of the silly putty? Let your imagination loose! You can probably think of many more interesting questions.

☞ *Example 4*

Did you think that our hydrogen peroxide experiment was really just a demonstration? You're right! Now, how can it be changed into an investigation that uses the scientific method?

Chemistry experiments involve two types of **variables**. Variables are things in an experiment that change. The **independent variable** is controlled and changed on purpose by the chemist who wants to see how the **dependent variable** changes as a result. The chemist decides how to change the independent variable before the experiment starts. Changes in the dependent variable are measured during the experiment.

When a small piece of potato is added to a cup of hydrogen peroxide, a reaction occurs that releases heat energy. The release of energy raises the temperature, which can be measured with a thermometer. The experiment now has a dependent variable that can be measured—the change in temperature.

One independent variable to investigate is the **concentration** of the hydrogen peroxide. For example, if the hydrogen peroxide is diluted by mixing it with an equal amount of water, will the temperature change be half as much? Chemists often use a **serial dilution** to study the effect of different concentrations on chemical reactions.

The steps in making a serial dilution of hydrogen peroxide are shown in the following diagram. Each medicine cup labelled one through four contains half as much hydrogen peroxide as the cup before it. The fifth cup contains only water.

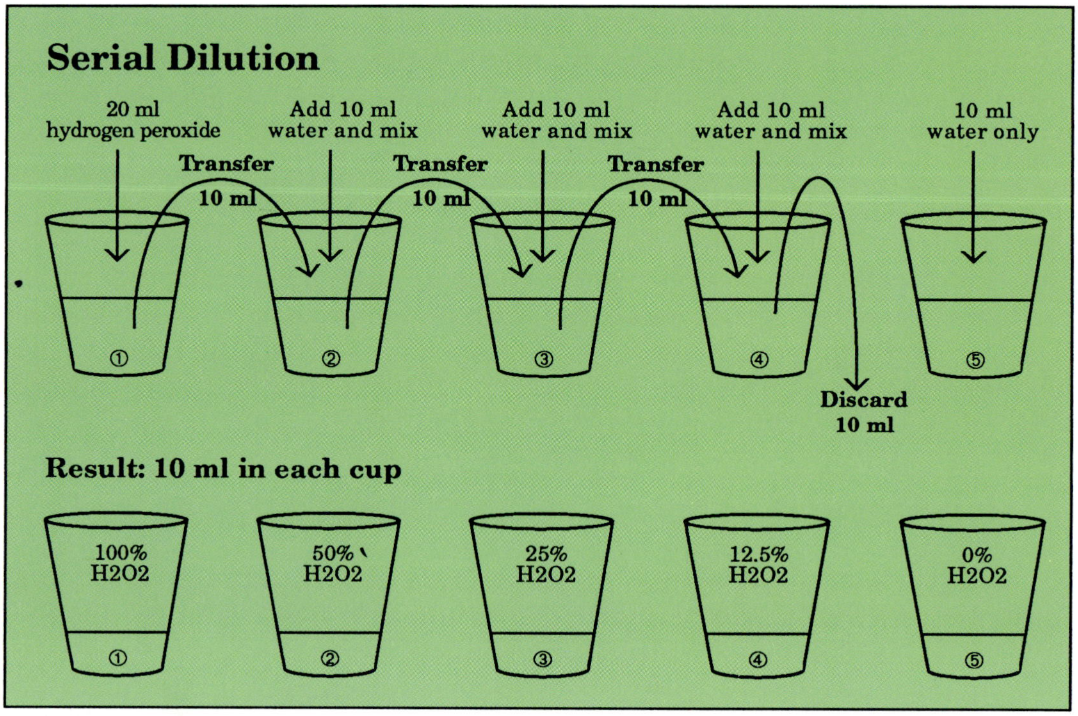

The concentration of hydrogen peroxide (independent variable) in the first four cups has been changed in a definite pattern.

The hypothesis for this investigation might now include the prediction that the temperature rise (dependent variable) due to the release of energy might also change in a definite pattern.

The three cups in which the hydrogen peroxide is diluted with water are the **experimental group**. A well-designed experiment also needs a **control group**. The fifth cup contains only water. The first cup contains only hydrogen peroxide. These two cups are the control group. Temperatures measured in the cups of the experimental group are compared to the temperatures measured in the control cups.

If you would like to try doing a serial dilution, you can get plastic medicine cups at a pharmacy. The cups are handy because they are marked for volume.

It's important to make careful measurements.

Thinking about the variables you can change and measure in an experiment will help you develop an idea for a chemistry project that is an investigation. Thinking about how the experimental group will compare to the control group will help you write a hypothesis for your investigation.

Stating Your Purpose

Once you have chosen a chemistry problem you want to study, you are ready to write a clear statement of the purpose of the experiment you plan to do:

"The purpose of this experiment is to determine if diluting the reactant, hydrogen peroxide, will decrease the amount of energy released during the chemical reaction."

You can also state the purpose in the form of a question:

"Does decreasing the concentration of the reactant, hydrogen peroxide, result in a decrease in the amount of energy released during the chemical reaction?"

Each of these sentences states the purpose clearly. They also make a prediction about the results. If defined clearly, the statement will also indicate the kind of measurements that will be made and what data will be collected.

Soon you will be ready to begin your lab notebook. A statement of the purpose of the investigation will be one of the first entries. A more detailed explanation of the lab notebook can be found in Chapter 3.

Gathering Information for Your Project

Your search for information should begin in a library. This is where you will find books, encyclopedias, magazines, and newspapers that contain articles about chemistry. You will find that librarians are eager to help you locate things about your topic. Be sure to ask for their help.

You can use the card catalog or on-line computer catalog to find out what books the library has. In the subject catalog, look under the headings "Chemistry," "Science Experiments," or other key words suggested by the ideas you are considering. If you have trouble, the librarian can help you find useful books, perhaps from other libraries.

The Reader's Guide to Periodical Literature will help you find magazines that contain articles about chemistry. Use the same subject headings to find a list of articles. Each entry will show the name and volume of the magazine, as well as the title and page numbers of the article. If there are articles you want to read, you will probably have to ask the librarian to get the magazines for you.

Are you wondering if scientists really go to the library before starting a new project? The answer is a definite "yes." Many laboratories and chemical companies have their own special chemistry libraries. Their librarians not only help the chemists find information in their own library, but also help locate information that may be in other libraries around the world.

Since you can't expect to remember everything you read, you will need to make some notes while you are at the library. Write your notes on 3" x 5" index cards. Later, you can select the most useful cards and arrange them in the best order. Each card should contain notes from only one source. For magazines, start by writing down the name of the author, then the title of the article, the name of the magazine, the volume number, and the date of publication. For books, write down the author's name, the book title, the publisher, and the date and place of publication. As you read, record the page numbers and write out your notes. If you need more than one card, repeat the author's name only and continue your notes—be sure to number the order of the extra cards.

If you copy the exact words from a published work, mark the copied section of your notes by enclosing it in quotation marks. If you use this information in your display or project report, you must rewrite it using your own words. Copying is called **plagiarism**; it is dishonest and illegal.

Use quotation marks when you copy information word-for-word.

People in your community are another source of information. Chemists work as teachers in high schools, colleges, and vocational schools. Hospitals, water treatment plants, and laboratories may have employees trained in chemistry. For some chemistry topics, a doctor or veterinarian may be helpful. Asking for the help of a professional chemist may seem scary, especially if you are shy and don't know the person. However, don't worry. Just introduce yourself and briefly explain your idea. You will probably be surprised at how friendly and willing the chemist will be.

Most chemists will be happy to share their knowledge.

CHAPTER 3

Planning Your Project

Once you know what chemistry problem you want to investigate, the next step is to decide how you will find the answers. In a scientific investigation, the answers to a problem come from the experiments that are done. An experimental design describes what materials are needed and how they will be used.

More About Experimental Design

In Chapter 2 you learned how a chemist changes the independent variable to see how the dependent variable will change. Perhaps you have already thought that there are other factors which could change the outcome of the hydrogen peroxide experiment. If so, you are right! A well-designed experiment must be done under **controlled conditions**. One condition, the concentration of hydrogen peroxide, is changed on purpose. A second condition, the temperature during the reaction, is measured. All other conditions must not be allowed to change at all.

To make sure that the temperature changes are due to the different concentrations of hydrogen peroxide, several other things must be controlled. Since temperature will be measured, plastic medicine cups may be a poor choice for containers; they may lose heat energy through the thin walls. Styrofoam cups are better **insulators**. All the cups should be the same size. At the beginning of the experiment, the liquid in each cup should be at the same temperature. The potato slices should be cut from a single potato. They should be cut to the same size and shape. You may think of other things that must be the same for all groups.

Everything in an experiment must be carefully controlled – even potato slices!

A well-designed experiment must also make sure that the data collected really mean something. Suppose you try the hydrogen peroxide experiment and there is no pattern in the temperature differences from cup to cup. Does this lack of a pattern really mean the hypothesis is incorrect? Could there have been an error in making the serial dilutions? Could the bottle of hydrogen peroxide be too old? There is no way to be sure.

To eliminate this problem, the experiment should be repeated several times. Each repetition is called a **trial**. How many trials a chemist will conduct depends on the data being collected. In some cases, there may be only a few trials. In other cases, there may be hundreds or more. In most cases, five trials will usually make sure that useful data is collected. You can always increase the number of trials if you need to.

If you find a chemist to help you, ask that person to check your experimental design. Ask for advice about changes that will make the experiment safer and more practical. Ask the chemist to look for things that need to be controlled and how to control them. If no chemist is helping you, you must have a responsible adult look at your design. Perhaps your teacher or a parent can review your plan and at least make sure it is safe.

Lab Notebook

Everything about a scientific investigation is written in a lab notebook. You will need a spiral-bound notebook or a three-ring binder that you use only for your science project. Loose pieces of paper can get lost, so write everything in the lab notebook.

Lab notebooks help to keep your investigation organized.

The first page should have the project's title and your name. Your teacher or the science fair rules may require that you also include additional information on the first page. A table of contents on the next page will help readers find each part of your project.

As you learned in Chapter 2, the first entry in your lab notebook should state the purpose of your investigation. Next, transfer the useful information from your note cards to your lab notebook. Be sure to include all the information about the sources of the information. Notes from interviews with chemists should also be in your lab notebook. There should be a complete list of the materials you chose and a detailed description of how you will use them. All the measurements and observations you make during the experiment need to be carefully recorded in the notebook.

Your notebook doesn't have to be perfectly neat, but your handwriting should be clear enough for other people to read. Your notebook will be a part of your science fair display. Judges will look at it.

☞ *Materials*

A list of materials must be exact. It identifies what—and how much—will be needed to do the project. You may wish to put the list in alphabetical order. The example below shows a list of materials for the hydrogen peroxide experiment.

<u>Materials</u>

graduated cylinder	styrofoam coffee cups, 25
hydrogen peroxide, 100 ml	thermometer
knife	water, 200 ml
potato	

For some projects, chemicals may have to be ordered from a supply company. If you need to order something, ask your teacher or a parent to place the order. You will need to allow several weeks for delivery of your order. Several companies that sell chemicals are listed at the end of this book.

☞ *Procedure*

A detailed description of each of the steps in the experiment is called the **procedure**. List the steps in the order in which they are done. Your procedure should be written in your lab notebook. It must be approved by an adult, like the chemist you asked to help you.

You may find a **flow chart** useful. It is a simple version of your procedure that will remind you of the steps and their correct order. A flow chart is very useful during the experiment. You may not have the time to search through your procedure when you need to know what to do next. The sample flow chart is for the hydrogen peroxide experiment.

Flow Chart

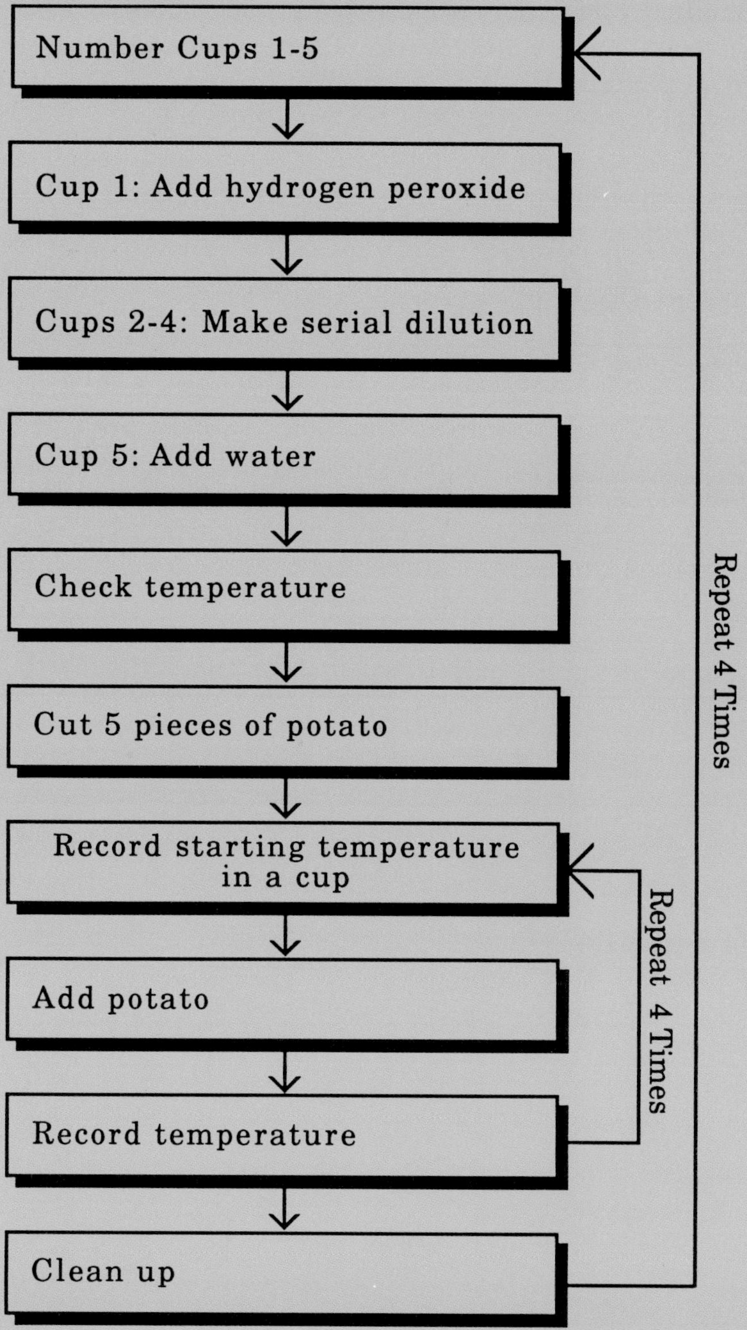

Number Cups 1-5
Cup 1: Add hydrogen peroxide
Cups 2-4: Make serial dilution
Cup 5: Add water
Check temperature
Cut 5 pieces of potato
Record starting temperature in a cup
Add potato
Record temperature
Clean up
Repeat 4 Times
Repeat 4 Times

Time Schedule

If you start planning your project several months before it is due, you will have time to enjoy your work. There will be time to correct errors and redesign experiments. You may find yourself doing a whole series of related experiments—and that makes science fun!

You will need time to decide on a topic, do library research, gather materials, plan and conduct the experiment, analyze the data, write a report, and make a display. Estimate how much time each step will take, then add a little more time! Allow plenty of time for the delivery of chemicals if you must order something.

<u>Sample Time Schedule for a Chemistry Investigation</u>

Week 1: Choose a problem
Week 2: Plan the experiment and gather materials
(This may take more than one week.)
Week 3: Conduct the first trial of the experiment
Week 4: Analyze the data; redesign the experiment if needed
Week 5: Repeat additional trials of the experiment
Week 6: Finish collecting and analyzing data
Week 7: Make the display
Week 8: Practice the oral report; present your project

CHAPTER 4

Doing Your Project

Safety

Safety is always the first concern in every step of designing and doing a chemistry project. As described in Chapter 2, one of the first things you should do is find a chemist who will help you. If you don't have the help of a chemist, you must at least have an adult who will review your experimental design for possible dangers. You need an adult to act as your supervisor.

You will need a place where you can work on your project. The best place would be in the lab of your helpful chemist. If that isn't possible, a table in your classroom or space on the kitchen counter at home can be used. Maybe someone can help you set up a table in your garage, storage room, basement, or in your own room. When you are picking a place for your work, think about your needs for getting water, for mixing chemicals, and for cleaning up after your experiment. You will also need space to properly store the chemicals and other materials when you are not using them.

It's smart to wear gloves when using chemicals.

You will need a rubber apron and safety goggles to wear when you are using chemicals. Some chemicals may also require that you use rubber gloves to protect your hands. Be careful what you touch while you are using chemicals. Keep your hands away from your face and out of your mouth.

Keep all chemicals in their original containers. Store each one according to the instructions on the label. Everything you use should be clearly labelled. Before you begin an experiment, find out the proper way to clean up any spills. Be sure that all the clean-up equipment you might need is close to your work area.

If you will need electricity, try to find battery-powered equipment. It's usually safer and easier to use. Keep all cords and wires away from water and chemicals. Remember that electrical equipment can get hot!

You must mix chemicals and conduct experiments only in the presence of your supervisor. Your supervisor doesn't have to stand over you and watch every move you make. However, your supervisor should be close enough to see or hear what you are doing. Even when you are repeating your experiment for additional trials, your supervisor must be present. Even if you know what you are doing, your safety, and the safety of others, requires that you never do chemistry experiments alone.

Keeping a Record of Your Results

After you have designed your experiment—but before you actually begin doing it—you will need to decide how you will record the data. What measurements and observations should you record? There are two kinds of observations. A **qualitative observation** does not use measurements—it uses words. The following statements are qualitative observations:

> The reaction in cup 1 produced the most bubbles.
> No bubbles were produced in cup 5.
> The temperature in cup 4 was lower than in cup 3.

A **quantitative** observation does use measurements—it uses numbers. The following statements are quantitative observations:

> The starting temperature in each cup was 25°C.
> The highest temperature reached in cup 2 was 33°C.
> The temperature reached in cup 1 was 7° higher than that reached in cup 2.

Quantitative observations are more precise. They provide more accurate information to use in evaluating the hypothesis. Plan to record as many quantitative observations as possible.

All observations are written in the lab notebook. You can't write down too much!

☞ *Data Charts*

Using a **data chart** will make recording observations easier and faster. In a data chart, the independent and dependent variables are identified. The values for the independent variable are filled in before the experiment. The values for the dependent variable are filled in as the experiment proceeds. You can't expect to remember everything you saw and measured during an experiment. The data chart should be ready to fill in before the experiment begins. A page in the lab notebook for recording data during the hydrogen peroxide experiment might look like the sample below.

DATA

Trial Number _________

	Temperature (°C)	
Cup Number	Starting	Highest
1		
2		
3		
4		
5		

Additional data could be collected during each trial by adding another independent variable that could be measured for each cup—the time. Instead of just watching for the highest temperature reached during the reaction, a timer could be used to help you record the temperature every 15 seconds during the reaction. Then there would be a record of all the energy changes during the entire reaction. The data chart would have to be changed. There would be more data to fill in. The new data chart might look like the one below.

DATA

Trial Number ________

	Temperature (°C)				
Time (s)	Cup 1	Cup 2	Cup 3	Cup 4	Cup 5
0					
15					
30					
45					
60					
75					
90					
105					
120					
135					
150					
165					
180					

Five of these charts would be necessary—one for each of five trials.

Scientists must be **unbiased** when observing what happens in an experiment. They record exactly what they see and measure, even if the data do not match their predictions. Remember, with the scientific method you can't get "wrong" results.

If your observations are not what you expected, look for a reason. Maybe your hypothesis was not correct. Then you can state a new one that better explains your observations. Maybe you discovered an uncontrolled variable in your experiment. Then you can redesign the experiment and do it again. Whatever the results, you will always learn something of value from an experiment.

☞ *Metric Measurements*

Scientists all over the world make measurements using the metric system. No matter what language they speak, they all understand and use the same metric units for length, volume, mass, and temperature. You should use the metric system, too.

Scientists around the world use metric tools for measuring.

A meter stick or metric ruler is used to measure length. A **meter**, the basic metric unit of length, is divided into smaller units, **centimeters** and **millimeters**. Just as 100 cents equal one dollar, 100 centimeters equal one meter. Millimeters are very small—1000 millimeters equal one meter. Select the unit that best measures the variable. You would use centimeters to measure the distance a polymer sample flows, but you might use meters to measure how high it bounces.

Chemists use a **graduated cylinder** to measure volume. A graduated cylinder is much like a measuring cup, only it is more accurate. A **liter**, the basic metric unit of volume, is divided into **milliliters**. One thousand milliliters equals one liter. Unless a large volume is needed, most chemicals are measured in milliliters.

A **balance** is a tool used to measure mass. The basic metric unit is the **gram**. A dollar bill has a mass of about one gram. The mass of heavy objects is measured in **kilograms**.

A thermometer is used to measure temperature. It is marked in degrees Celsius (C) for metric measurement. Water boils at 100°C and freezes at 0°C. Body temperature is 37°C and room temperature is about 20°C.

Analyzing Results

Sometimes an experiment yields so much data that it's hard to see what they mean. There are several methods scientists use to help figure out data. First, when an experiment has several trials, an average for the trials can be calculated. This reduces the amount of data. For example, suppose you have recorded the highest temperature reached during the reaction in cup 1 for each of five trials. You can find the average high temperature. Add together the temperatures for each trial; then, divide the sum by 5 (the number of trials). The answer is the average high temperature for cup 1. After the averages for the remaining cups have been calculated, it will be easier to compare the experimental and control groups.

Data		Calculating an average	
24°C	24°C	**26.8°C**	The average high temperature during the reaction in cup 1 is **26.8°C.**
28°C	28°C	5) 134	
27°C	27°C	100	
26°C	26°C	34	
29°C	29°C	30	
	134°C	40	
		40	
		0	

Second, scientists use graphs to help them see patterns in the data. The graph most frequently used by chemists is a **line graph**. In a line graph there are two axes—one vertical and one horizontal. The **horizontal axis** shows the independent variable. The **vertical axis** shows the dependent variable. Each axis is labelled with the name of the variable, the units used to measure it, and a scale of number values. The graph has a title that tells what observations are being shown. After the data are plotted, the points are connected with a line. In order to compare several sets of data, several lines can be plotted together. Each one is identified by a different color or pattern. Below is a line graph and a data chart from which the data were taken to make the graph.

Change in Temperature During the Reaction of Hydrogen Peroxide

Cup 1

Time (s)	Temp (°0)
0	22
15	22
30	23
45	23
60	25
75	28
90	32
105	36
120	39
135	42
150	41
165	38
180	37

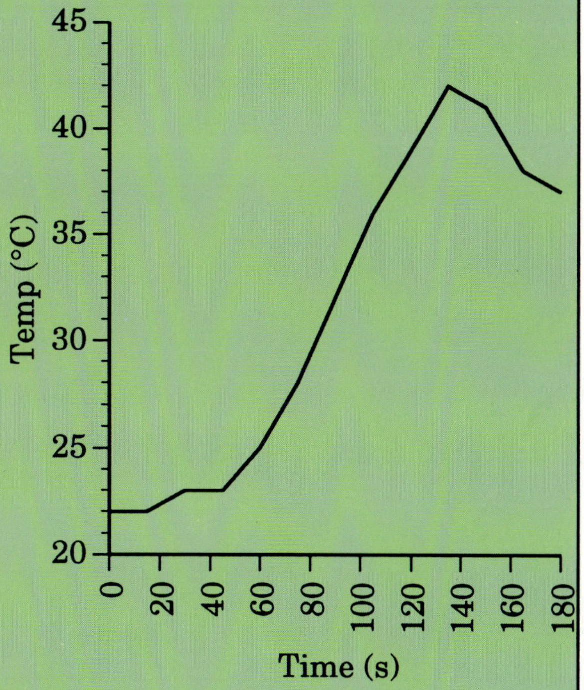

There are several other types of graphs. They include circle graphs, pictographs, and bar graphs. These types are mostly used to display data. They are not usually useful in analyzing data. More information about these graphs can be found in Chapter 5.

Drawing Conclusions

When all the data have been collected and analyzed, conclusions are drawn. Conclusions tell what the data mean. In writing conclusions, a scientist looks at the hypothesis. Do the data support the hypothesis? What new experiments might further test the hypothesis? If the data do not support the hypothesis, what new hypothesis does the data suggest? What new problems can be identified? Write all your conclusions in your lab notebook. Try to explain how you reached all of your conclusions.

Go back over the notes in your lab notebook. If you think you made mistakes, write down what you think they might be. Explain how you think these mistakes may have changed your observations. Remember, there is no such thing as being "wrong."

Another important part of your conclusions is an explanation of how the results of your experiment might be useful to other people. For example, suppose you did an experiment to find out what an acid does to various building materials. In your conclusions you could discuss which materials would best resist acid rain. Science fair judges are very interested in what uses you can think of for your results.

CHAPTER 5

Presenting Your Project

The last step of the scientific method is to tell others about the experiment and its results. Other scientists need to know what work has been done. Scientists tell about their work in many ways. One way is to write an article for a magazine or a book. Another way is to give a speech at a scientific meeting. When you take part in a science fair, you can tell others about your experiment, too.

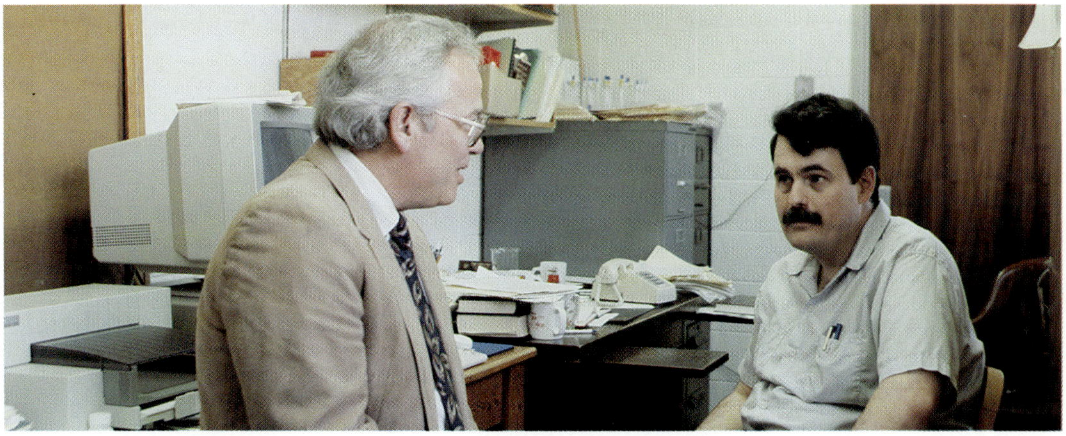

Part of a scientist's job is to tell others about their work.

Writing a Report

Most science fair judges expect to see a written report, not just a lab notebook. Some fairs have rules about what to include, as well as the order. Before you begin writing, find out what the local rules are.

Written reports can be typed or printed by computer. If you write neatly, handwriting is alright, too. Put the pages in a folder or a three-ring binder.

The first page, called the title page, should have the title of your project and your name. Your teacher might ask you to also write the date, the school's name, and your grade. The second page should be a table of contents to help readers find the parts of your report.

The remaining parts of the written report follow the same order as the lab notebook. The lab notebook is the source of all the information in the report. The following list shows what usually needs to be included:

Contents for Written Report

Purpose	Data
Background Research	Conclusions
Hypothesis	Bibliography
Materials	Acknowledgements
Procedure	

The statement of purpose tells what you wanted to find out in your chemistry project. In the background research section, give the information you found in books and magazines, and tell what you learned when you talked with chemists and other experts. In the hypothesis section, explain your hypothesis and experimental design.

A list of materials, the procedure that tells how you used them, and the data you collected and observed are in the next sections. You should include any line graphs that you made to help you analyze your data. You may also want to include other types of graphs that display your data. These graphs are described in the next section of this book.

Using your lab notebook as a guide, state your conclusions. Discuss possible errors and describe new problems suggested by your results. Explain how your results can be applied in our world.

A bibliography lists all the sources of information you used in writing your report. The entries are in alphabetical order by the author's last name. You can find examples of bibliographies in the books and magazines you used. Your science textbook may have a bibliography. Two examples follow to help you write a bibliography:

For a magazine:

Barinaga, Marcia. "The secret of saltiness." Science 254:664-665 (1991).

For a book:

Hodges, Laurent. Environmental pollution. New York: Holt, Rinehart and Winston, 1977.

In the final section, you should thank everyone who helped you with your project. Name each person and tell how that person helped.

Making Graphs to Show Your Data

Chemists often graph the data collected during an experiment to better understand what the data mean. They most often use line graphs. Making a line graph to analyze data is explained in Chapter 4.

Graphs are also used to display data because they pack a lot of information into a small space.

When you want to compare several groups you can use a **bar graph**. In this kind of graph, bars are drawn to represent the average measured for each group. Bar graphs are a lot like line graphs. The horizontal axis is used to represent the independent variable, and the vertical axis is used to represent the dependent variable.

Drawing a bar graph takes careful planning. The height of each bar must represent the average for one group. You must decide how high to make the tallest bar. All the bars should be the same width. Look at the following data chart and bar graph as an example.

Bar Graph

Highest Temperature Measured during Hydrogen Peroxide Reaction

Concentration of hydrogen peroxide (%)	Average high temperature (°C)
100	42
50	34
25	30
12.5	28
0	25

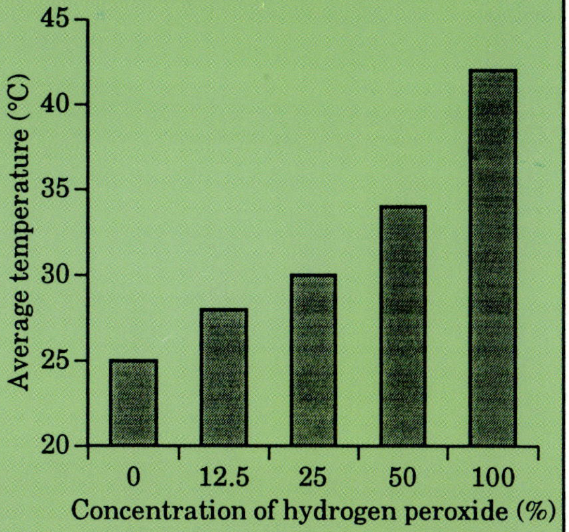

A **circle graph** is another way to display information. It is the type of graph to use when you have several parts that make up a whole. For example, in a project on polymers, you might find out how plastic made in the United States is used for certain purposes. A large circle represents all the plastic made. The circle is then divided into parts. The size of a slice of the circle represents how much of the total is used for one purpose. That slice is labelled with its use. A circle graph is shown below.

Circle Graph

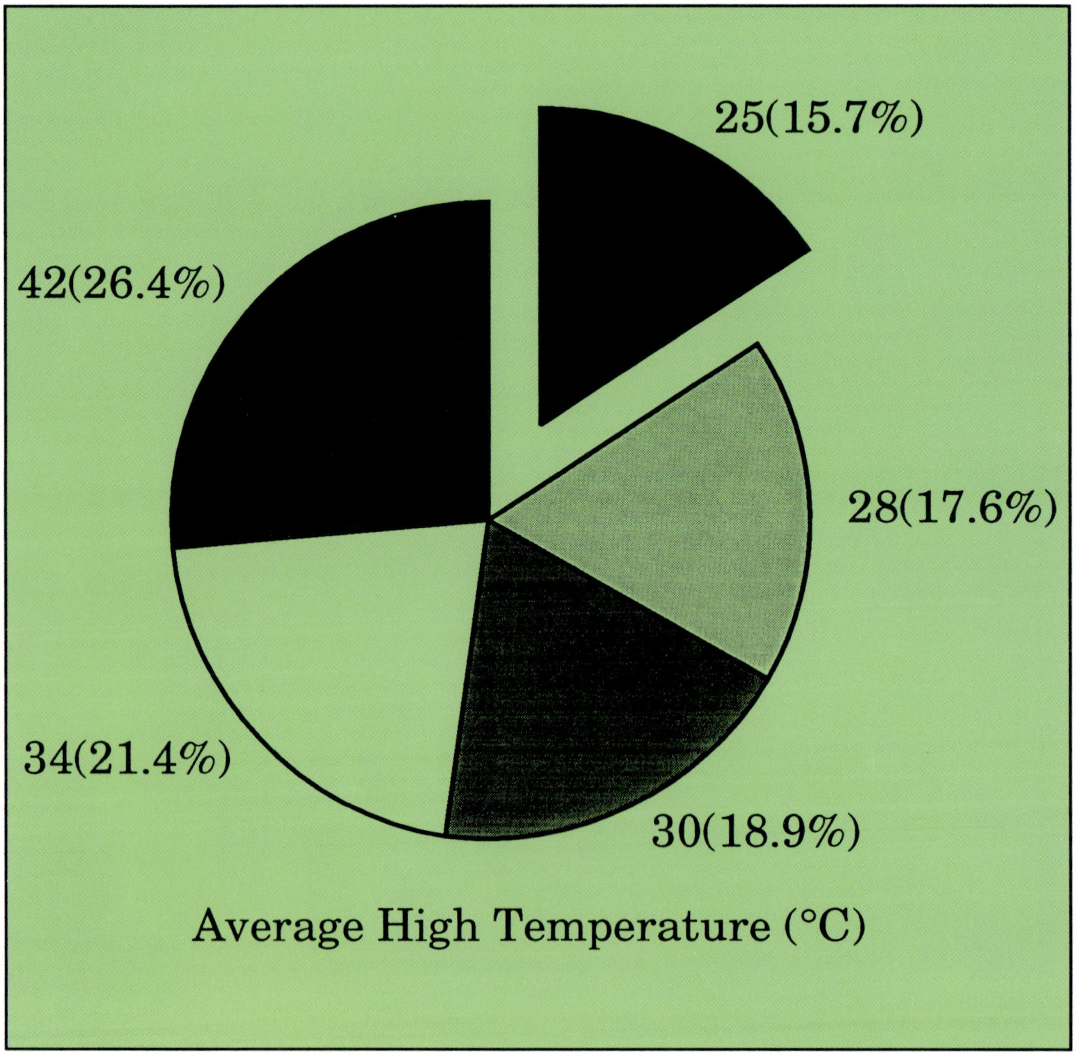

A **pictograph** is constructed like a bar graph, except that it uses a repeating symbol to represent the amount. If there is a fraction in the amount, only a part of the symbol is used. Pick a symbol that fits your project. Take some time to plan a pictograph's layout. Decide how many symbols you will need in each row. You will use many copies of the symbol, and all of them in the pictograph should look the same. You could carefully draw your design and then make many copies on a copy machine. You could also cut out the symbol from a small piece of cardboard and then trace around it to make the rows.

Pictograph

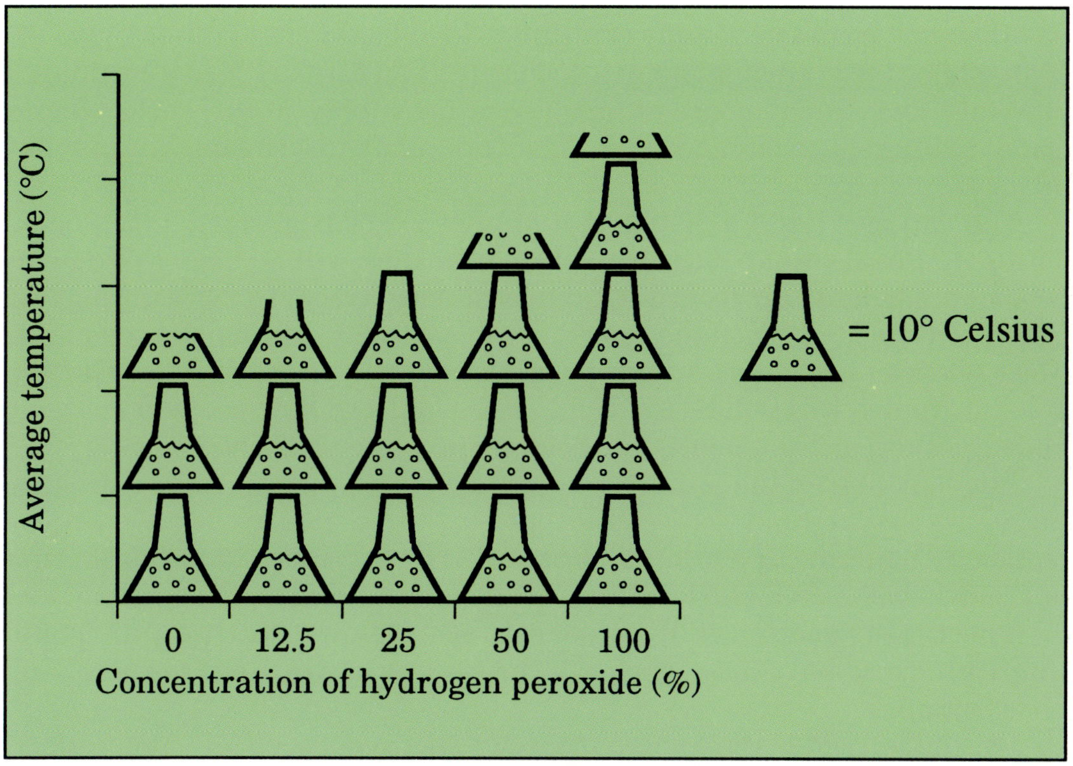

All graphs must be clearly labelled so the reader will know what the numbers mean. Each graph must have a title that tells what information is being shown. Lettering should be neat. You can buy press-on or stick-on letters at an office supply store. Choose a size that fits the graphs you have planned. You can also trace around letter stencils. If you can use a computer, a graphing program will arrange and print the graphs for you.

If you make the graphs yourself, you might want to use graph paper. Graph paper has lines printed on it that divide the page into squares. These squares can be used as guidelines in planning a graph. You can buy graph paper at an office or school supply store.

Organizing Your Display

When you visit a science fair, what do you notice first? Usually a good looking display catches your eye. The lettering is easy to read. The drawings, photos, and graphs are neat and colorful. The display tells the story of the project.

The most common display is a three-sided board that stands by itself. It should be large enough, but not too large. You can buy heavy cardboard or foam-filled display board at an office or art supply store. These boards come in different colors and sizes. They are ready to write on, tape on, pin on, or staple on.

You can also build a display out of plywood or peg board and hinges. These materials make sturdy displays, but they will be heavy. You can even cut up a large cardboard box into a three-sided board for your display. You might need to cover the brown cardboard with white paper or paint. Most science fairs have limited space for each project. Check the rules to see that your board will be the right size. Consider how you will get your board to class or the science fair before you make your choice.

Put your title on the center section of the board. Choose letters that are large and easy to read. You can use two- or three-inch press-on letters or a stencil. You can cut out letters from colored art paper. Plan the spacing of the title before you start putting on any letters. Check the spelling, too!

The center section will be seen first. Below the title, put some interesting graphs of data, photos of the results, or drawings to show your observations.

On the left side, state the purpose or the hypothesis of your project. Include a brief summary of your experimental design. You might want to use your flow chart instead of a step-by-step procedure. Photographs of you doing some of the steps will add interest. The photo will also show the judges that you did the experiment yourself.

On the right side, show any data that didn't fit on the center section.

Briefly state your conclusions. If you have room, you can include some ideas for new experiments.

Before you attach any of the pieces, place them on the board to see how they look. Move them around until you find a pleasing, simple arrangement. Have someone check your printing for correct spelling, punctuation, and grammar. Make sure everything is labelled correctly.

Finally, you are ready to put the display together. Select a glue that doesn't wrinkle your paper, use double-stick tape, or put loops of tape on

Some science fairs have hundreds of displays.

A good looking display will earn more points from the judges.

the back of the paper. Try to avoid using staples.

Stand your display up and see how it looks. Place your written report and lab notebook on the table in front of the display. The rest of the table space can also be used. You could put out samples of the materials you used or the products of your experiment. If you do put objects on the table, be sure they do not cover up the lower part of your display.

Talking to the Judges or to Your Class

You will probably be expected to talk about your project to science fair judges or to your class. In four or five minutes you will have to do the following:

- ★ Introduce yourself.
- ★ Give the title of your project.
- ★ State the purpose of your project.
- ★ Tell why you chose this project.
- ★ Explain what you did.
- ★ Show your results.
- ★ Explain what you learned.
- ★ State your conclusions.
- ★ Ask if there are any questions.

It's a good idea to use notes to help you remember what you want to say. Plan your speech and practice it in front of other people. Ask your practice audience for helpful ideas. After practicing several times, put away your notes. Try using just your display to help you remember. If you still feel better with notes, use them—but try not to just read from them.

When you are talking to the judges or to your class, stand to the side of your display. Point out parts of the display as you talk about them. Use your written report to show details to the judges. Talk slowly so you can think about what you are saying. If you get nervous or confused, take a deep breath and pause. Then go on with your explanation.

When you are finished, ask your listeners if they have any questions. If you don't know the answer, just say you don't know. If you don't understand a question, ask the person to repeat it. Before you answer, you might want to state the question yourself. Then the judge will know you heard the question correctly.

Try to be relaxed when you talk about your project.

Judging of Your Project

The science fair judges who visit you at your display will check your project for an award. Each science fair usually has its own judging sheets. However, most judges will be checking these things:

★ Scientific thought—The judges will see if your project follows the scientific method. They will look for a clearly stated problem and proper procedure, including the use of controls. They will look for variables and how they were controlled. They will check that the conclusions are supported by the results. The judges will look for evidence that you did your own experiment.

★ Creative ability—The judges will want to know how you got your idea. This is the score that will be low if you did an exhibit or demonstration, or if someone else did the actual work. If your idea came from a book like this one, explain how you used your imagination to add to the idea. More points will usually be given for scientific thought and creative ability than for the other areas.

★ Understanding—Your display should show what you learned. Judges will ask questions to see if you understand the key scientific ideas in your project. They will want to know what background sources you read and if you prepared a bibliography. They will look at your display and written report to see if they tell the complete story.

★ Clarity—The judges will score projects higher if the problem, experimental design, data, and conclusions are clearly stated. The written report and display should be easy to read and understand, even for the average person.

★ Technical skill—The judges will try to decide if you are able to do the experiment and make the measurements yourself. They will want to know who helped you and how. They will look at the graphs in your written report and on your display to see if they are done correctly. Judges want to see that you put thought and effort into the construction of your display.

Final Words of Encouragement

Doing a science fair project takes time and hard work. When you finish your project, take pride in what you have done. You have not been "acting" like a scientist—you have been a scientist! You used the scientific method to investigate a chemistry problem. You now know how to ask a question and find out the answer. You have discovered the excitement and wonder of doing science.

Where do you go from here? Maybe you'll want to enter your project in some other competition. Perhaps you'll want to do another project for next year's science fair. As you get better at using the scientific method, you can develop projects to enter in more advanced contests. The Westinghouse Science Talent Search, the Thomas Edison/Max McGraw Scholarship

Program, or the Duracell/NSTA Scholarship Competition are some possible goals for the future.

Let your imagination lead the way!

When you receive an award, take pride in it!

Companies That Sell Scientific Supplies

Carolina Biological Supply Co.
2700 York Road
Burlington, NC 27215
1-800-334-5551 (East of the Rockies)
1-800-547-1733 (Rockies and West)
1-800-632-1231 (North Carolina)

Connecticut Valley Biological Supply Co., Inc.
P.O. Box 326
Southampton, MA 01073
1-800-628-7748 (U.S.)
1-800-282-7757 (Mass.)

Edmund Scientific Co. (colored filters)
101 E. Gloucester Pike
Barrington, NH 08007-1380
1-609-573-6250

Nasco
P.O. Box 901
Fort Atkinson, WI 53538-0901
1-800-558-9595

Ward's
P.O. Box 92912
Rochester, NY 14692-9012
1-800-962-2660

Science Competitions

The Duracell/NSTA Scholarship Competitions
1742 Connecticut Avenue, N.W.
Washington, DC 20009

Edison/McGraw Scholarship
Dr. Kenneth R. Roy, NSSA
National Director, P.O. Box 380057
East Hartford, CT 06118

Westinghouse Science Talent Search
Science Service
1719 N. Street, N.W.
Washington, DC 20077-6532

GLOSSARY

acid - a compound containing hydrogen that reacts with a base to form a salt.

balance - a tool used to measure mass.

base - a compound that reacts with an acid to form a salt.

buffer - a compound that modifies the effect of an acid or base.

chemical bond - a force that holds atoms together in molecules.

chemical reaction - chemical change.

concentration - strength.

conclusion - what you interpret the results of an experiment to mean.

control group - a group in an experiment in which as many variables as possible are kept constant because they could affect the outcome of the experiment.

controlled conditions - making sure that anything that changes during an experiment is done on purpose.

data - the observations and measurements that you make in an experiment.

dependent variable - the factor or condition that changes as a result of the presence of, or a deliberate change you make in, the independent variable.

DNA - a large molecule that contains the code of life that offspring inherit from their parents.

experimental design - the plans you make so you can do an experiment. The design includes what you will use and how you intend to use them.

experimental group - a group in which all variables are the same as those in the control group *except* for the factor that you are following in your experiment.

flow chart - a list that writes out a shortened version of the steps you want to follow in doing your experiment. As you complete each step, you should check it off the list.

graduated cylinder - a tool used to measure the volume of a liquid.

hypothesis - a statement that gives a possible answer to a question, sometimes called an "educated guess." To see if it is true or not, a hypothesis is tested by doing an experiment.

horizontal axis - the line at the bottom of a graph that represents the independent variable.

independent variable - the factor or condition that you want to study. In an experiment, you intentionally change this factor.

inorganic matter - materials that come from minerals rather than living things.

insulator - a material that is a poor conductor of heat energy.

mass - the amount of matter, or "stuff," that is present. Weight is often confused with mass. Weight is the pulling force of gravity on matter.

organic matter - materials that come from living things.

pH scale - scale from 0 to 14 used to tell the strength of an acid or base.

plagiarism - copying word-for-word what someone else has written and not giving credit to that person.

polymer - a large molecule with a repeating structure.

product - a new substance formed during a chemical change.

qualitative observation - an outcome of an experiment that is not an amount that can be measured, such as color.

quantitative observation - an outcome of an experiment that is measurable, such as numbers of individuals.

procedure - the steps you follow in an experiment.

reactant - the original substance that will be changed during a chemical reaction.

saturated solution - a solution having the greatest possible concentration.

scientific method - a systematic strategy scientists use to discover answers to questions about the world. It includes making a hypothesis, testing the hypothesis with experiments, collecting and analyzing the results, and arriving at a conclusion.

serial dilution - a method of reducing the strength of a liquid that results in a pattern.

trial - a test that is one of a group of repeated tests.

unbiased - not allowing your preferences to interfere with collecting or analyzing data in an experiment.

variable - some factor in an experiment that can be changed.

vertical axis - the line on the left side of a graph that represents the dependent variables.

INDEX